Boston Terrier
Collectibles

Donna S. Baker and Paul Hiller

Schiffer
Publishing Ltd

4880 Lower Valley Road, Atglen, PA 19310 USA

Dedications

To my wife and co-collector, Joan, who is an inspiration to me in every way and whose knowledge and forbearance have made collecting, and this book, possible.

—PH

To Karen, Peter, and Adam, my three wonderful "littermates" — this one's for you, with all my love and gratitude.

—DSB

Library of Congress Cataloging-in-Publication Data

Baker, Donna S.
    Boston terrier collectibles / Donna S. Baker and Paul Hiller.
        p. cm.
    ISBN 0-7643-1884-5 (pbk.)
    1. Boston terrier--Collectibles--Catalogs. I. Hiller, Paul,
1935- II. Title.
NK4891.2 .B34 2003
688.1--dc21

2003009580

Designed by Mark David Bowyer
Type set in Windsor BT/Souvenir Lt BT

ISBN: 0-7643-1884-5
Printed in China
1 2 3 4

Published by Schiffer Publishing Ltd.
4880 Lower Valley Road
Atglen, PA 19310
Phone: (610) 593-1777; Fax: (610) 593-2002
E-mail: Info@schifferbooks.com
Please visit our web site catalog at
www.schifferbooks.com
We are always looking for people to write books on new and related subjects. If you have an idea for a book, please contact us at the above address.

This book may be purchased from the publisher.
Include $3.95 for shipping.
Please try your bookstore first.
You may write for a free catalog.

In Europe, Schiffer books are distributed by
Bushwood Books
6 Marksbury Avenue
Kew Gardens
Surrey TW9 4JF England
Phone: 44 (0) 20 8392 8585
Fax: 44 (0) 20 8392 9876
E-mail: Bushwd@aol.com
Free postage in the UK. Europe: air mail at cost.

Contents

Acknowledgments

Thanks are very much in order to Dotty Truman, a Boston Terrier lover and longtime collector from Virginia, whose support and assistance made this book a reality. Dotty's energy, enthusiasm, and willingness to open her home to a tangle of photography equipment are all very much appreciated! Thanks as well to Bruce Waters, for helping with the photography, and to the Schiffer design team for making this book look beautiful.

And of course, our thanks to Tivoli, Tara, Bits, and Tyler, for their inspiration, loyalty, good humor, and unconditional affection.

Introduction

If you've ever shared your life and home with a dog, you know the unparalleled joy that comes from being greeted each day by a pair of bright, adoring eyes at one end and a cheerful, wagging tail at the other. From herding to hunting to guiding, dogs have served their human companions in many capacities throughout history. Some excel at chasing, some at sniffing, some at pulling, some at simply looking elegant. Regardless of breed or background, however, we cherish our dogs for their most universal and familiar role — protectors of the heart and nurturers of the soul.

True dog lovers are a passionate sort, prone to conversing at length about the particular virtues and qualities of their chosen breed or breeds. With an incredible diversity of size, shape, color, hair length, and personality among purebred dogs — not to mention the hardy and often unique characteristics of mixed breeds — it is certainly possible for each prospective dog owner to find just the right match of canine/human compatibility. And for those who favor a lively, inquisitive, and quite dapper looking companion, what breed could be more well-suited than the Boston Terrier, a dog described by authors Gerald and Loretta Hausman as "a mighty midget with the nicest face [and] the most compassionate and impressive eyes, set well apart in an elegant, Herculean head."

Fondly known as "the American Gentleman," the Boston Terrier traces its origins to the late nineteenth century and resulted from a cross between English Bulldogs and English Terriers. One of the breed's forefathers was a dog named Hooper's Judge, said to be mostly Bulldog and owned by Robert C. Hooper of Boston, Massachusetts. Hooper's Judge was bred to a female known as Burnett's Gyp; their descendents eventually led to the modern day Boston Terrier. Other crosses involved the French Bulldog, which had its beginnings from the same sources as the English Bulldog. The Boston Terrier is one of very few and probably the most prominent of breeds originating in the United States, and is therefore considered a truly American breed.

The Boston Terrier Club of America was formed in 1891, and two years later the American Kennel Club (AKC) recognized the Boston Terrier as a breed. The first Boston to be registered with the AKC was a dog named Hector, while the first Boston to achieve AKC champion status was a female named Topsy — she won her championship at the Philadelphia Dog Show in April of 1896.

With its compact build and easygoing demeanor, the Boston Terrier moved forward very quickly in popularity. From 1920 to 1963, Bostons ranked among the top ten breeds registered with the AKC and many celebrities of the era, such as Joan Crawford, Charles Boyer, and Douglas Fairbanks, Jr., owned one of these charming, spirited little dogs. Drawings of Boston Terriers done by Robert Dickey, a highly esteemed commercial artist, appeared regularly in the *Saturday Evening Post* as well as other popular magazines, helping to increase the visibility and fame of the breed. So enamored of Bostons were dog fanciers in the state of Massachusetts that they began a movement to have the Boston Terrier become their official state dog. The breed was officially elevated to that position in 1979.

Today, the Boston Terrier continues to maintain a favorable position near the top half of all breeds registered annually with the AKC. They are certainly in first place, however, to all who have been lucky enough to share their life with a much-loved Boston as companion, confidante, and four-legged best friend.

The Collectible Boston Terrier

Once your life becomes enriched by one or more canine companions, it's not much of a stretch to become captivated with canine images on all sorts of antiques and collectibles. Indeed, what better accoutrements for the dog lover than a collection of items bearing the same cherished countenance as the real life dog or dogs in the home? As author Laurence Sheehan remarks in *Living With Dogs*, "dogs settle into a home pretty much the way people do, gravitating to a favorite corner, window, or chair. Like children, they leave their toys scattered around — a rubber squeaky here or a half-gnawed bone there. These objects, along with the leashes, collars, food bowls, dog beds, and all the other paraphernalia associated with keeping a dog, turn a house into a habitat…When dog owners add dog art and collectibles to their decor, whether specific to their favorite breed or in praise of dogs in general, the habitat becomes a gallery."

Among collectors of canine memorabilia, Boston Terrier related items rank very high — indeed, many feel that they are number one in popularity in this regard. Accordingly, this book pays homage to the many delightful antiques and collectibles portraying or honoring Bostons in a host of different ways. Given that for many categories there is hardly a finite number of items available, it is not meant to be a complete look at this subject, but rather a pictorial survey illustrating the wide range of collectible pieces. In some cases, we have noted that the items pictured are representative of many others that can be sought and found by avid collectors and enthusiasts.

As noted earlier, the Boston Terrier and the French Bulldog evolved from similar ancestors, so it is not surprising that some pieces can be interpreted as depicting either of these breeds. In those cases where features of both breeds are evident, we have noted "Boston Terrier or French Bulldog" in the caption. While some collectors may want to limit their acquisitions to those items that are undeniably one breed or the other, we feel the occasional ambiguity simply reflects the shared origin of the two breeds and does not detract from the appeal or worth of these items.

The quest for Boston Terrier collectibles is a wonderful challenge, in that many of these items are scarce and difficult to find, but they are not — as with some breeds — non-existent, and there are certainly enough around to make such a quest successful. And should you decide to fill your home or just one shelf with the results of your quest, you will be in good company — there are a great many avid collectors all over the country, as enthusiastic about Boston Terrier collectibles as they are about the breed itself.

About the Values

The values listed in this book are intended to provide readers with a general idea of what they might expect to pay for the same or similar item in today's market. The values represent a guideline only and are not meant to "set" prices in any way. It is entirely possible to purchase an item for a higher or lower amount than the value shown here, as many factors affect the actual price paid. These factors include condition, scarcity, the location of the market (i.e., antique shop, dog show, flea market, live auction, Internet auction, etc.) and the buyer's relative desire to own a particular item.

Chapter 1

Artwork

"Jealousy," print by artist Eda
Doench, for Guttman and
Guttman, c. 1922, 14" x 19".
$300-$500.

"Lady Be Good," original etching
by artist Morgan Dennis (1892-
1960), 7.5" x 5.5". Born in Boston,
Massachusetts, Dennis is well known
for his dog illustrations, which
appeared in many magazines and
books. Today, his works are very
popular with collectors of canine art
and memorabilia. $250-350.

Boston Terrier and cat etching, also by Morgan Dennis,
7" x 9". $250-350.

Small color print of two Boston puppies, unknown artist, 8" x 10". $25-50.

Framed print by artist Robert Dickey (1861-1944), "The Family Album," 7" x 9.5". Bostons also appear in Dickey's popular *Mr. and Mrs. Beans* book and comic strip (see pages 18-19). $10-20.

"A Social Error," original etching of two Boston puppies drinking, by Marguerite Kirmse (1885-1954), 8.5" x 10.5". Born and raised in England, Kirmse lived in America during the 1930s. In Paul's collection are several other Kirmse etchings of Bostons, including "The Rivals," "Expectation," "The Sting," and "Beans." Along with "A Social Error," these date from the 1930s and 1940s and are valued at $500-1,200 each, depending on size.

Three panel wooden
frame with richly
colored prints of
Bostons in different
positions, c. 1950s,
12" x 26.5". $200-300.

Signed and numbered print
by artist and illustrator Roy
Anderson, "The Yankee
Dandy," one of 450 issued,
14" x 17". The four stamps
shown below the print were
designed by Anderson and
issued on September 7,
1984, to commemorate the
100th anniversary of the
American Kennel Club. The
stamps feature eight dog
breeds, including a Boston
Terrier at top left. $250-350.

Richly colored print of Boston Terrier sitting outside, wooden frame has collar decoration at top and gold label at bottom, c. 1950s, 25" x 17". $200-300.

Pen and ink drawing, "Ladies First," by European artist Zito, c. 1960s-1970s, 9" x 12". Other drawings with Bostons by this same artist include "Just a Gigolo" and "Hey, What's On Your Mind." $100-200 each.

Ladies first

Chapter 2
Bookends

Bradley & Hubbard cast iron bookends, standing Boston Terriers on rectangular bases, 5" h. x 4.75" w. $400-600.

Reverse side of base, showing B & H insignia.

Differently colored version of the Bradley & Hubbard bookends. $400-600.

Cast iron bookends, one set painted, both made by the Hubley Manufacturing
Co., of Lancaster, Pennsylvania, c. 1930s, 4.75" h. x 5.5" w. $200-350 each set.

Hubley sticker on underside of the painted
bookend. It is rare to find such a sticker.

Painted cast iron bookends with puppy-like seated
Bostons, probably by Hubley, c. 1920s, 6" h. $300.

Cast iron bookends, black and white Bostons standing on green bases, 5" h. x 5.75" w. $250-350.

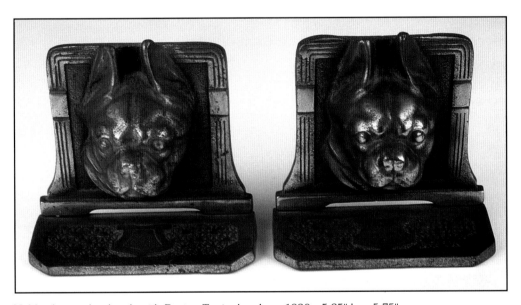

Hubley bronze bookends with Boston Terrier heads, c. 1930s, 5.25" h. x 5.75" w. A painted version of these bookends also exists. $500-600.

Bronze bookends with Boston Terrier heads, "Y" at top stands for Yale University (which used Bostons as their mascot interchangeably with Bulldogs), c. 1930s, 4" h. x 4.5" w. $400-600.

These bookends are the same as the previous set with Boston heads, but do not have the "Y" at top. $400-600.

Bronze bookends, Bostons with doghouses, c. 1940s, 6.5" h. x 4.75" w. $200-300.

Metal bookends, mother Boston with two pups, 4" h. x 4" w. $75-150.

Brass bookends, strumming minstrels with Bostons at their feet, c. 1940s-1950s, 6" h. x 4" w. $300-500.

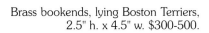

Brass bookends, lying Boston Terriers, 2.5" h. x 4.5" w. $300-500.

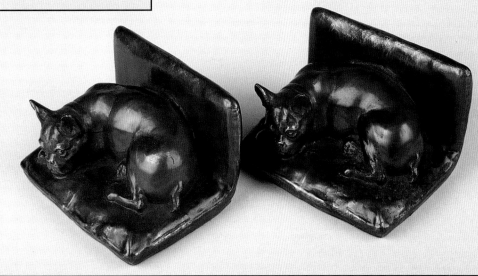

Bronze or bronze coated bookends, three Bostons on rock wall, made by Jennings Brothers, 6.75" h. x 5.25" w. $400-500.

Bookends, cast iron figurines mounted on wooden bases, diamond shaped label on bottom for "Littco Products," 5" h. x 5" w. $75-150.

Onyx bookends with metal Boston figurines, 4" h. x 3.25" w. $200-300.

Ceramic bookends, seated Bostons peering over white and yellow bases, 4" h. x 2.25" w. $100-200.

Ceramic bookends, Boston heads on yellow bases, Japanese, c. 1970s, 4.5" h. x 3.5" w. $75-175.

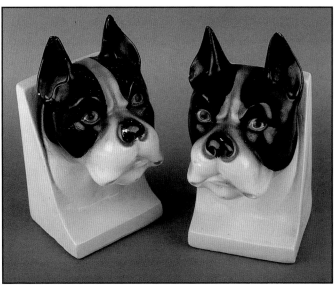

Ceramic bookends, slightly different style Boston heads on smaller yellow bases, also Japanese, 4.5" h. x 3" w. $75-175.

Chapter 3

Books and Ephemera

These four books with Boston Terrier themes are examples of the many works of fiction featuring "The American Gentleman." From left: *Here, Tricks, Here*, by Lebbeus Mitchell, Cupples and Leon Publishers, New York, 1923; *White Boots*, by Helen Orr Watson, Houghton Mifflin Company, Boston, 1948; *Skitch, The Message of the Roses*, by Vincent G. Perry, Denlinger's, Fairfax, Virginia, 1975; *Mr. And Mrs. Beans*, by Robert L. Dickey, Frederick A. Stokes Co., New York, 1928. *Mr. and Mrs. Beans* (with dust jacket, as shown), $400-600; others, $40-60 each.

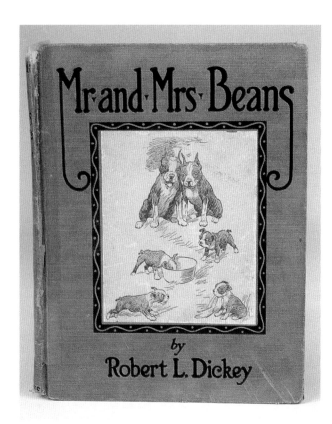

Another copy of Robert Dickey's *Mr. and Mrs. Beans*, this one without the dust jacket. One of Robert Dickey's prints featuring Bostons can be seen on page 8. $200-350 without dust jacket.

There are many additional books, magazine covers, and sheet music titles featuring Boston Terriers — those shown here are just a sampling of the wonderful variety available. A few additional examples are listed below.

Additional books:
Dogs & Puppies, by Frances Trego Montgomery (Borse and Hopkins, 1908)
Tim, by Ethelbert Talbot (Harper & Bros., 1914)
Pat and Pal, by Harriet Lummis Smith (L. C. Page & Co., 1928)
Mi TripWest Bi Me, by Sever Freeman (privately printed, 1930)
50 Favorite Songs, by Mary Nancy Graham (Whitman Publishing Co., 1935)
Little John of New England, by Madeline Brandeis (Grosset & Dunlop, 1936)
Young MacDonald Had a Farm, by Carlyle Leech and John McKenna (Steven Daye, Inc., 1944)
The Playful Little Dog, by Jean Horton Berg (Wonder Books, Inc., 1959)
Patch the Pup, by Dennis Twomey and Catherine Goudreau (Gannett Books, 1990)

Additional sheet music titles:
"I'm on the Jury"
"Just a Kid"
The Smart Set"
"War-Time Tommy"
"No Vacancy"
"I Like the Hat, I Like the Dress"
"Come Out of the Kitchen, Mary Ann"

Front and back cover of *Mr. and Mrs. Beans* comic book by Robert Dickey, © 1939, United Features Syndicate, Inc. Front copy reads "Children's World Macy's Toyland, New York World's Fair"; inside cover reads: "Robert Dickey, Artist and Creator of Mr. and Mrs. Beans, 'The Man Who Makes Dogs Talk!'" Rare, hard to find. $300-500.

Detail from back cover of *Mr. and Mrs. Beans* comic book.

Cover of *Capper's Farmer* from October 1931, featuring three nervous Boston pups with rooster. $10-30.

Magazine cover from *St. Nicholas*, with Boston Terrier looking to meet a new friend. A magazine for children, *St. Nicholas* was published from 1873 to 1939. $10-30.

Humorous magazine cover from *Nature Magazine*, November 1931. $10-30.

Magazine cover from *Liberty*, Dec 3, 1932. $10-30.

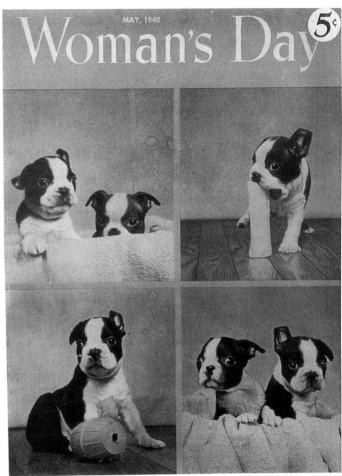

A bevy of Boston pups grace this *Woman's Day* cover from May 1948. $10-30.

Cover of *The Canine Collector's Companion*, January/February 1986, with illustrations by Robert Dickey. $10-30.

Sheet music, "When I Woke Up This Morning," Will Rossiter, The Chicago Publisher, copyright MCMXI. $25-40.

Sheet music, "Bring Me Back My Lovin' Honey Boy," Will Rossiter, The Chicago Publisher, copyright MCMXIII. $25-40.

Sheet music, "When the Moon is Shining," published by Jos. W. Stern Co., 1920s. $50-100.

Sheet music, "When You're All Dressed Up and Have No Place to Go," Will Rossiter, The Chicago Publisher. $50-100.

Sheet music, "Gimme a Bit," Herman Darewski Music Publishing Co. Ltd., London. $50-100.

Assortment of postcards, c.1900 to present, all featuring a Boston Terrier with his or her human companion. These are just a selection of the many postcards available to collectors, many featuring Bostons alone with no people in the image. Values range widely, but average $30-50 for the better ones.

Chapter 4

Clocks and Lamps

German Black Forest clock in the shape of a
Boston Terrier with clock face in eyes, wooden,
c. 1930s, 7" h. $500-700.

Wooden clock in the shape of a Boston
Terrier head, also with clock face in eyes,
American, made by Sessions Co., c. 1930s,
6" h. $500-600.

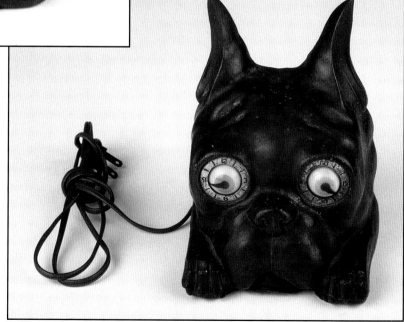

Graniteware enamel clock, German,
c. 1920s, 4" h. Rare. $600-900.

Bavarian perfume lamp, seated porcelain Boston
Terrier in center, c. 1920s, figurine is 8" h. Rare.
$500+

Table lamp with Boston Terrier base, base and part of dog
are flocked, decals on shade may have been applied later,
15" h. $200-300.

Lamp with Boston Terrier sitting on oval base, marked "Made in Czechoslovakia" underneath base, figurine is 5" h. x 6" w. $500+

Chapter 5

Desk Accessories

English pewter inkwell with old style
Boston head on top, c. 1910, 3" h.
$600-800.

Bronze inkwell, very heavy, head
lifts up, 3.5" h. $500+

Glass inkwell, Boston pup with metal hinge at
collar, Czechoslovakian, c. 1930s, 3.75" h.
$400-600.

Figural inkwell of tree stump with Boston Terrier
standing beside it, 1.25" h. $150-250.

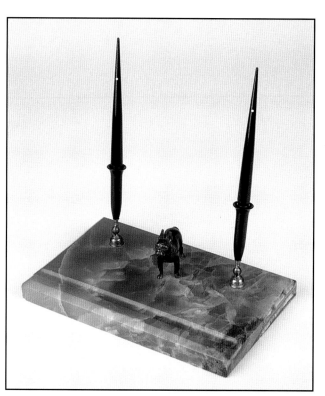

Desk set, Vienna bronze figurine of Boston Terrier on green marble
base, 2.5" h. to top of dog x 9.5" w. $800-1,000.

Bronze blotter and inkwell (in stump) with Boston Terrier or French Bulldog figurines in various positions, inkwell is 9" l. x 3" w.; blotter is 3.5" h. x 4.5" w. Rare. $1,000-1,500 for both.

Vienna bronze desk blotter, standing Boston on top, 2" h. x 4" l. x 2" w. $600-800.

Two Vienna bronze seals, c. 1940s, 3" and 2.5" h. Rare. $1,200-1,500 for both.

Bradley & Hubbard business card holder with Boston Terrier heads on either side, metal with painted heads, c. 1920s, 3" h., base 2.75" x 2.75". $300-400.

Two stamp boxes and perpetual calendar, all sterling silver, c. 1930s-1940s. Calendar reads "Copyrighted 1893," has same Boston Terrier head as on one of the stamp boxes (except for color), and measures 2.5" x 2.25". Stamp boxes are 2" dia. each. $200-400 each.

Doorstops, Paperweights, and Banks

Doorstops

Right facing cast iron doorstop with original chain and doorknob ring, made by the Hubley Manufacturing Co., of Lancaster, Pennsylvania, c. 1910, 10" h. x 6" l. $100-150.

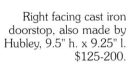

Right facing cast iron doorstop, also made by Hubley, 9.5" h. x 9.25" l. $125-200.

Two left facing cast iron doorstops, one brindle and one black and white, eyes rimmed in white, both made by Hubley, 9.5" h. x 9.25" l. each. Left facing doorstops are less common and harder to find than the right facing versions. $200-300 each.

Forward facing cast iron doorstop, molded on green collar with studs, small ring at rear of collar, stamped "CCO" underneath, weighs 6.75 lbs, c. 1920s-1930s, 9.5" h. x 9" l. Rare. $250-350.

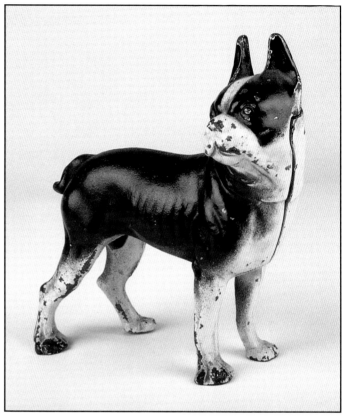

Right facing cast iron doorstop, cutout areas underneath in both front and back, 8" h. x 7.5" l. $125-200.

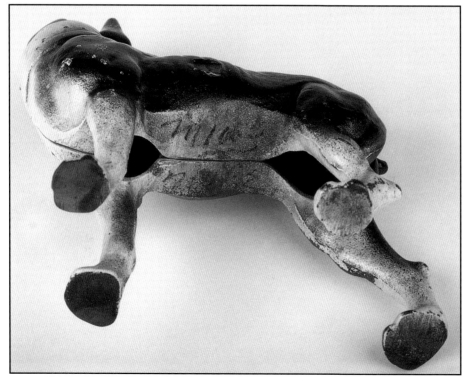

Side view of above doorstop, showing the cutout areas underneath. The presence of such cutout areas usually, but not always, indicates that the doorstop is original. Nearly all reproduction doorstops are solid underneath and do not have cutout areas; however, some original doorstops are also solid underneath.

Two cast iron doorstops in unusual sitting position, one black and white, one brown and white, c. 1920s, 6.5" h. x 6" l. $300-450 each.

Two cast iron begging Boston doorstops, red eyes and lips on both, comprised of two pieces joined at mid body by a single slot screw that ends in a square nut on right side, 8.5" h. x 2.75" w. Hard to find. $600-900.

Right facing cast iron doorstop, head turned further to right than usual, wide white rims on eyes, solid underneath, 8.5" h. x 7.5" l. $200-300.

Left facing cast iron doorstop, black and white, "Skoog" on collar, 7" h. x 7" l. Hard to find. $450-650.

Cast iron doorstop, standard style but with very unusual addition of blanket over back with letters "BSA" for Boy Scouts of America, c. 1920s, 10" h. Very rare. $1,000+

Detail of face and collar on the "Skoog" doorstop.

Cast iron doorstop, possibly by Hubley, c. 1920s, unusual height of 7". Rare. $1,000+

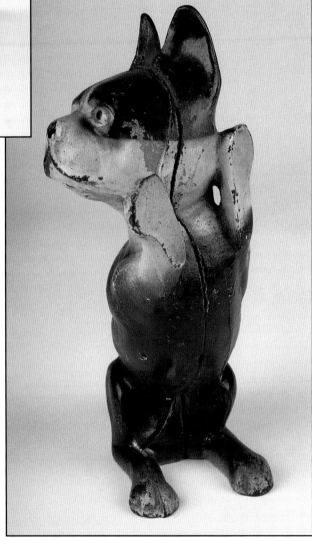

Cast iron one sided doorstop with small base, no wedge, sitting puppy facing right, c. 1930s, 7" h. x 6" w. $400-600.

Cast iron doorstop or screen holder, brown and white Boston Terrier with paws up, also comes in black and white, c. 1920s, 9.5" h. Note oval opening that separates the dog's left leg and head — this will be found on a vintage piece but not on a reproduction. $600-800.

Very heavy cast iron doorstop, flat back on body only (not head), seated Boston puppy with red collar and "Fido" in red on the front, c. 1930s, 5" h. $150-200.

Cast iron mechanical doorstop, puppy with bow around neck and bone by paws that moves, c. 1930s, 5.25" h. x 3" l. $500-800.

Cast iron one sided doorstop with wedge by Bradley & Hubbard, black and white Boston standing on grassy green base, 9.5" h. x 9.75" l. $500-600.

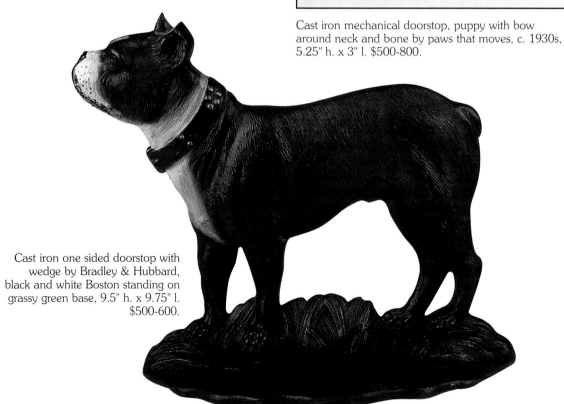

Cast iron one sided doorstop with wedge, whimsical features, back reads "Copyright 1927, A. M. Greenblatt, Studio 17," 9.5" h. x 4.5" w. $400-600.

Two unusual cast iron flat sided doorstops, one enameled in brown and one black with painted green eyes, both have cast letter "A" on reverse side, 11.5" h. x 5" w. $600-800 each.

Cast iron sitting puppy doorstop, one sided wedge, c. 1920s, 7" h. x 6.5" l. $400-600.

Cast iron doorstop, portrays Ally Sloper (an English cartoon character) with Boston Terrier at his feet, c. 1920s-1930s, 10.25" h. Rare. $1,000-1,500.

Cast iron doorstop, very different style from most and in quite good condition, possibly German, c. 1920s-1930s, 14" high. Rare. $1,000+

Cast iron doorstop, one sided with wedges, Boston Terrier holding red, white, and blue ball between paws, 7.5" h. x 11" l. Rare. $1,000+

Right facing bronze doorstop on marble base, 9.5" h. x 9.5" l. Rare, probably one-of-a-kind. $1,000+

Antiqued bronze doorstop or figurine, marked "JB 2472" underneath, may stand for Jennings Brothers, c. 1930s, 7" h. x 7.75" l. $400-600.

Paperweights

Two Hubley cast iron figurines or paperweights, c. 1930s, seated dog is 4" h. x 5.5" l., standing dog is 4" h. x 5.25" l. $300-400 each.

Hubley cast iron paperweight, c. 1930s, 4" h. x 5.5" l. $150-200.

Bronzed cast iron figurine or paperweight, also by Hubley, c. 1920s, 4" h. x 5.5" l. $150-300.

Assortment of smaller paperweights by Hubley, all cast iron with different finishes. 2.75" h. each. $75+ each.

Cast iron paperweight, brown and white standing Boston, 3" h. x 3" l. $75+

Cast iron figurine or paperweight, rare size, 4.5" h. x 4.5" l. $75+

Three metal paperweights, same standing Boston with different finishes, 2.5" h. x 2.5" w. $75+ each.

Small sterling paperweight, Boston Terrier or French Bulldog, inscribed on back "Boston," c. 1920s, 2" l. $300-400.

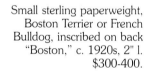

Small Tiffany bronze paperweight with d'oré finish, Boston Terrier or French Bulldog, similar to previous paperweight but does not have "Boston" on back, c. 1910-1920, 2" l. $600-800.

Banks

Right facing cast iron bank, also made as a paperweight, 5" h. x 4.75" l. $75-175.

Two right facing cast iron banks. Smaller one has opening on back and measures 4.5" x 4.5"; larger one has opening at rear of neck, was made by Vindex Toy Co. and has blue eyes (which would be a disqualification on a real Boston!), and measures 5.25" h. x 5.5" l. $75-175 each.

Cast iron bank made by the Hubley Manufacturing Co., of Lancaster, Pennsylvania, opening at rear of neck, 4" h. x 5.5" l. Note similarity of this bank to the three Hubley paperweights shown on page 42. $75-175.

Cast iron bank, facing forward with chain around neck, 5.5" h. x 6.75" l. $75-175.

Aluminum bank, made by the Vindex Toy Co., c. 1931, 5.25" h. x 5.5" l. $75-175.

Steel bank, possibly with brass coating, opening at rear of neck, 4.75" h. x 5.25" l. $75-175.

Chapter 7

Figurines and Sculpture

Cast Iron,
Bronze, and Other Metal

Brown and white cast iron figurine with chain around neck (chain may not be original), c. 1930s, 5" h. x 5.5" l. The same figurine made into a cigarette lighter can be seen on page 141. $250-450.

Cast iron figurine, all black, dealer from whom piece was purchased noted that it may have come from a foundry in Shippensburg, Pennsylvania and may have been made during a worker's lunch hour, 4.75" h. x 4.75" l. $40-55.

Bronze figurine of seated Boston on tasseled pillow, marble base, signed "F. Rieder," very heavy, 10.75" h. x 4.75" square base. Rare. $2,000+

French Art Deco sculpture of woman playing with Boston Terrier or French Bulldog, figures made of metal and some bronze, ivory ball, marble base, c. 1930s, 9" h., 14" x 4.5" base. Rare. $800+

Bronze figurine, Boston Terrier sitting on haunches and begging, 4" h. $150-300.

French Art Deco sculpture of woman reclining with Boston Terrier or French Bulldog at her side, made of gilt metal and bronze, woman's face and hands are French ivory, c. 1930s, 7" h., 17.5" x 5" base. Rare. $1,500+

Very heavy, finely detailed metal sculpture, Boston Terrier or French Bulldog, embossed mark on belly for Jennings Bros. (JB 2472), c. 1940s-1950s, 6.5" h. x 7" w. $300-450.

Metal reclining Boston Terrier or French Bulldog, also made by Jennings Bros. (marked with JB on bottom), hollow inside, 2.5" h. x 4" l. $75-125.

Bronze figurine of Boston Terrier jumping over marble wall, unusual piece, 3" h. x 6" w. Rare. $1,000+

Reverse side of jumping Boston figurine.

Bronze figurine on marble base, "The Bostonian" by American artist Chris Baldwin, c. 1980s-1990s, figurine 7" h., base 6.25" x 4.25". $500+

Bronze figurine of reclining Boston, also by Chris Baldwin, c. 1980s-1990s, 5.25" h. x 6" l. $500+

Vienna bronze figurine of Boston Terrier or French Bulldog in sitting position with head turned up, sculpted by Franz Bergmann (one of the most famous Vienna bronze sculptors), c. 1900, 11" h. Rare. $2,500-3,000.

Two reclining Boston figurines made of metal on wooden base, American artist, c. 1970s-1980s, 4" h., base 5.75" w. Rare, one-of-a-kind piece. $600-800.

Life-size, bronze figurine of pregnant Boston Terrier or French Bulldog, French artist, 20" l. Rare. $1,500+

Two bronze sculptures of Boston Terriers by American artist Kathleen O'Bryan Hedges, c. 1980s-1990s. From left: "After the Bath," 7" h.; "American Gentleman," 11" h. Born in Hawaii in 1948, Hedges is a self-taught artist who began sculpting dogs, chickens, horses, and other farm animals in 1984. A bronze plaque by this same artist is shown on page 122. $800-1,200 each.

Bronze figurine of puppy, a bell push also made by Gornik and signed on the side, c. 1920s, 2.25" h. x 3.25" w. Born in the late nineteenth century, Gornik studied sculpture at the Vienna School of Arts and Crafts and worked in both porcelain and bronze. Rare. $1,000+

Bronze figurine on rectangular base with bone bell push, made by Austrian artist Friedrich Gornik and signed on the side, "F. Gornik," c. 1920s, 3.5" h. Rare. $1,000+

Bronze figurine of standing Boston Terrier named "Stubbie," artist name Skoog (name of dog and artist on front of figurine), American, c. 1960s-1970s, 7" h. x 8.5" l. Rare. $800+

Vienna bronze figurine, standing Boston Terrier or French Bulldog with red collar and sad expression, 3" h. $300-500.

Very heavy Vienna bronze figurine, signed Bergmann, c. 1900, 5" h. x 6" l. Rare. $1,000+

Pair of Vienna bronze figurines of Boston Terriers or French Bulldogs, sculpted by Berman, another Vienna bronze sculptor, 2" h. x 2" l. $200-400 each.

Two small Vienna bronze figurines, also sculpted by Berman, sitting dog 1.5" h., reclining dog .75" h. $200-400 each.

Two Vienna bronze figurines of Boston Terriers or French Bulldogs, mostly white, one sitting and one standing, 1.5" h. each. $200-350 each.

Very heavy bronze figurine sculpted by American artist and dog enthusiast Trophy Frederick, limited edition of ten made in the likeness of one of her own dogs, contemporary, 12" h. $2,000-2,800.

Bronze figurine of standing Boston on rectangular base, made by the Gorham Co., c. 1920s, 4.5" h. x 4.5" w. $800+

Whimsical metal figurine of Boston playing the piano, red collar around neck, no markings, c. 1970s-1980s, 4" h. x approx. 5" w. $75-125.

Sterling silver figurine made by the Kirk Co., c. 1980s, 2" h. x 2.25" l. $300-400.

Porcelain and Ceramic

Three seated porcelain figurines made by Hutschenreuther, German, c. 1920s-1930s, largest one is 9" h. x 11" w., medium is 7.5" h. x 7" w., smallest is 6.25" h. x 6" w. Rare. $800-1,000+ each.

Detail of expression on largest Hutschenreuther puppy's face.

Hutschenreuther mark on bottom of puppies.

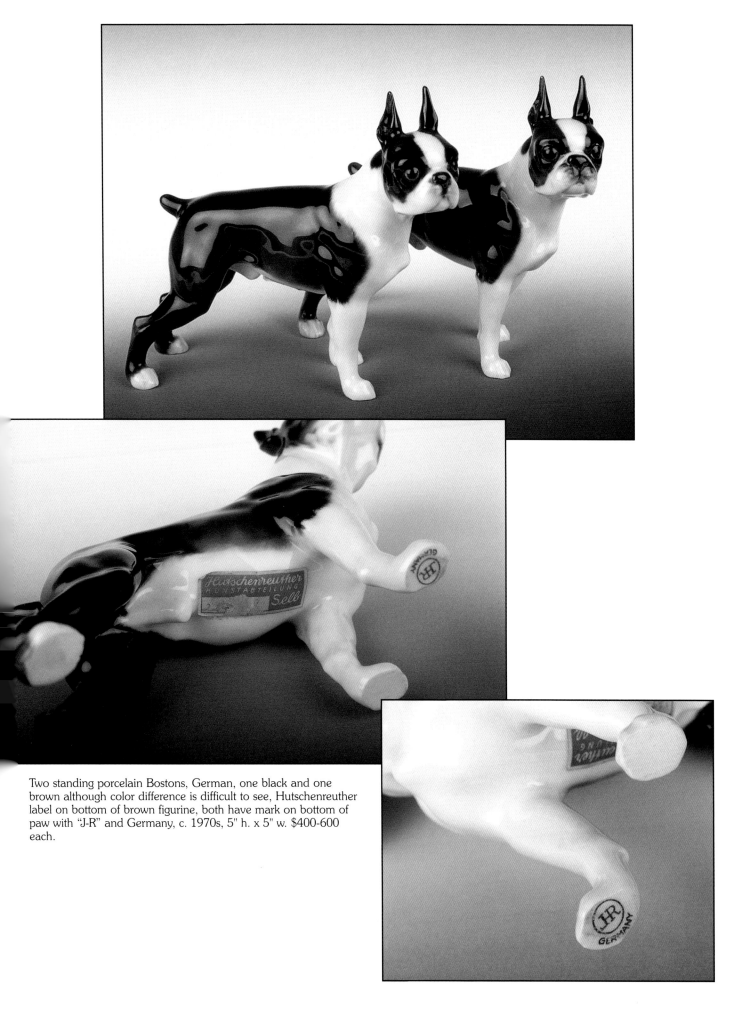

Two standing porcelain Bostons, German, one black and one brown although color difference is difficult to see, Hutschenreuther label on bottom of brown figurine, both have mark on bottom of paw with "J-R" and Germany, c. 1970s, 5" h. x 5" w. $400-600 each.

Two seated porcelain figurines by Meissen, one on left is older and more detailed, each is 6" h. x 4.5" w., German, both are marked on bottom with Meissen crossed swords in blue. Rare. $1,000+ each.

A similar porcelain figurine by Meissen, German, c. post 1950s, marked on bottom with Meissen crossed swords in blue, 5.75" h. x 3.75" w. Very rare. $2,000+

Two porcelain pups by Meissen chewing on their back legs, one marked on the bottom with Meissen crossed swords in blue, German, 5.5" h. each. Rare. $1,000+ each.

Meissen mark on bottom of black pup.

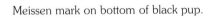

All white porcelain Boston Terrier or French Bulldog by Meissen, German, 6" h. x 7" w., Meissen mark on bottom. $600+

Very rare standing figurine by Meissen, German, c. 1900-1920, 6" h. x 7" l. $1,500+

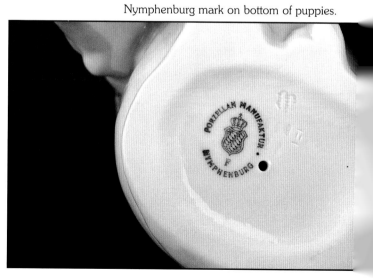

Nymphenburg mark on bottom of puppies.

Two very rare porcelain puppies by Nymphenburg, Boston Terriers or French Bulldogs, German, larger one is 6.25" h. x 4" w., smaller one is 4.75" h. x 3.25" w. Larger puppy, $800-1,000; smaller puppy, $600-800.

Rare porcelain figurine of Boston Terrier or French Bulldog reclining on pillow, made by Nymphenburg, beautiful detail and coloring, German, 7" h., base is 8" x 12". $1,500+

Another porcelain figurine of reclining Boston on pillow, also by Nymphenburg, German, c. 1950+, 7.5" h., base is 8" x 12". Very rare. $1,800+

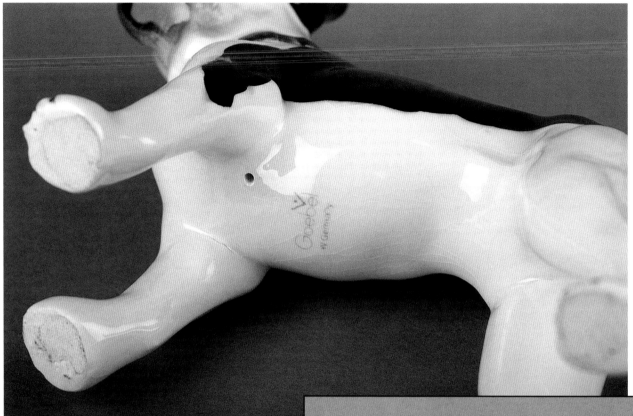

Mark on underside of Goebel figurine.

Porcelain figurine by Goebel, Boston Terrier or French Bulldog, German, c. 1930s, 4.25" h. x 4.5" l. $150-200.

Two porcelain figurines made by
Rosenthal, both marked on the bottom,
German, reclining dog is 4" h. x 9" l.,
seated dog is 6.5" h. x 5.5" w. $600-800
each.

Mark on bottom of
the reclining
Rosenthal figurine.

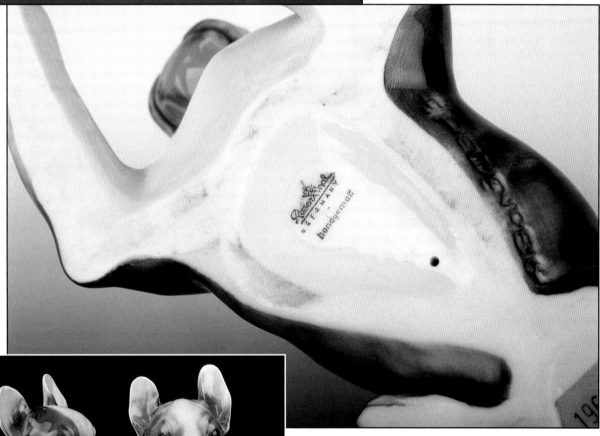

Two porcelain Boston puppies by
Rosenthal, German, c. 1930s, 7.5" h. x 6.5"
l. $700-900 each.

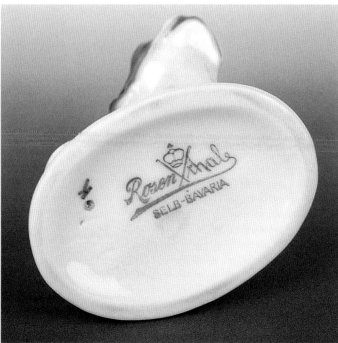

Tiny brown and white Boston puppy, also made by Rosenthal and marked on bottom, German, c. 1942, 2.5" h. A black and white version exists as well. $200-300.

Seated porcelain figurine by Gotha Pfeffer, German, c. 1930s, 6" h. x 5.5" w. $200-400.

Two reclining porcelain figurines, both German. Larger one made by Erphila, c. 1930s-1940s, 5" l. $150-250. Smaller one made by Gotha Pfeffer, c. 1930s-1940s, 3.5" l. Hard to find. $250-400.

Porcelain figurine by Metzler & Orloff, seated Boston in gray and white, German, c. 1930s, 5.5" h. x 5" l. $150-300.

Small porcelain figurine by Metzler & Orloff, German, c. 1930s, 2" h. x 1.75" l. $100-150.

Standing grey and white porcelain Boston wearing pink collar, also by Metzler & Orloff, German, c. 1930s, 4" h. x 4.5" w. $200-300.

Old Thuringian German porcelain figurine, seated Boston Terrier or French Bulldog, Muller & Co., c. 1920s, 6.5" h. x 5" l. $600-800.

Porcelain figurine of seated Boston Terrier, made by Plau von Schierholtz, German, c. 1930s, 5" h. x 4" l. Rare. $600-900+

Goldcrest mark on bottom of figurine at right.

Brown and white standing porcelain Boston, marked "Goldcrest Ceramics Corp." on bottom, 5.5" h. x 6" l. Goldcrest Ceramics was a later name for the Goldscheider company, originally based in Austria. $250-300.

Porcelain figurine of Boston Terrier or French Bulldog with dark ears (shown from the front and side), made by Heubach, German, 4" h. x 4.5" w. Rare. $600-800.

Bing & Grondahl porcelain figurine of Boston Terrier puppy scratching neck, sculpted by artist Jean Rene Gaugin, Danish, c. 1920s-1930s, 6.5" h. x 8" l. Rare. $1,000+

Two porcelain figurines by Bing & Grondahl, distinctive B&G underglaze mark and green hallmark, also marked 2330, Danish, c. 1955-1960, 7.5" h. $450-650 each.

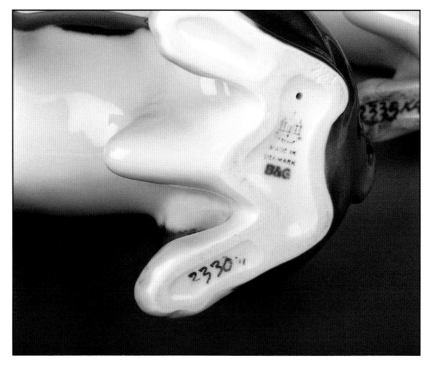

Mark on bottom of one of the Bing & Grondahl figurines.

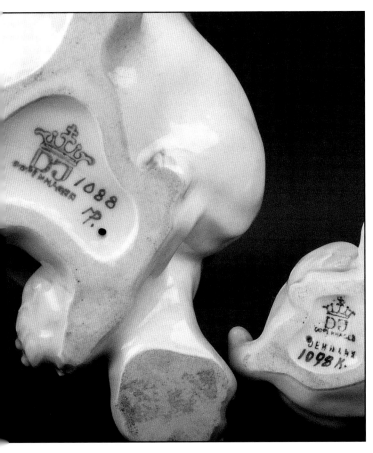

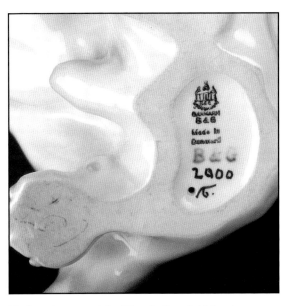

Mark on bottom of the Bing & Grondahl figurine below.

Marks on bottom of the Dahl Jensen figurines shown below.

Three porcelain figurines of Boston Terriers or French Bulldogs with similar styles and expressions, left and center pups by Dahl Jensen, right hand pup by Bing & Grondahl, all marked on bottom, Danish, c. 1930s. Larger Dahl Jensen pup is 6" h., $600-$1,000; smaller Dahl Jensen pup is 2.5" h., $200-300; Bing & Grondahl pup is 7" h., also comes in black and white, $1,000+.

Porcelain figurine by
Royal Copenhagen in
gray and white,
marked on bottom,
Danish, 4.5" h. x 5" w.
$300-400.

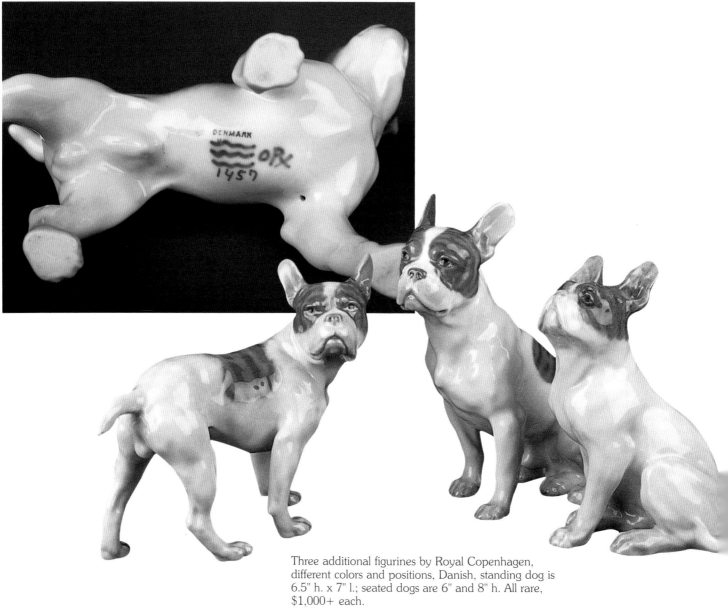

Three additional figurines by Royal Copenhagen,
different colors and positions, Danish, standing dog is
6.5" h. x 7" l.; seated dogs are 6" and 8" h. All rare,
$1,000+ each.

Very rare porcelain figurine of standing Boston by artist Felice Tosalli, Italian, c. 1930, 11" h. x 16" l. Known as a "Lenci Torino," a form of this piece is included the book *La Belleza DiLenci*, by A. Panzetta. $1,200-1,500.

All white Bing & Grondahl figurine, standing Boston Terrier or French Bulldog, Danish, 6.75" h. x 7" l. Rare. $600+

Sevres all white porcelain figurine, seated Boston Terrier or French Bulldog on rectangular base, French, c. 1920s, figurine 4" h., base 3.5" x 3". Very rare, museum quality piece. $2,500+

Large, all white seated porcelain figurine on oval base, unidentified mark on bottom, Russian, 8.75" h. $300-500.

All white porcelain figurine, Japanese, 6" h. $50-75.

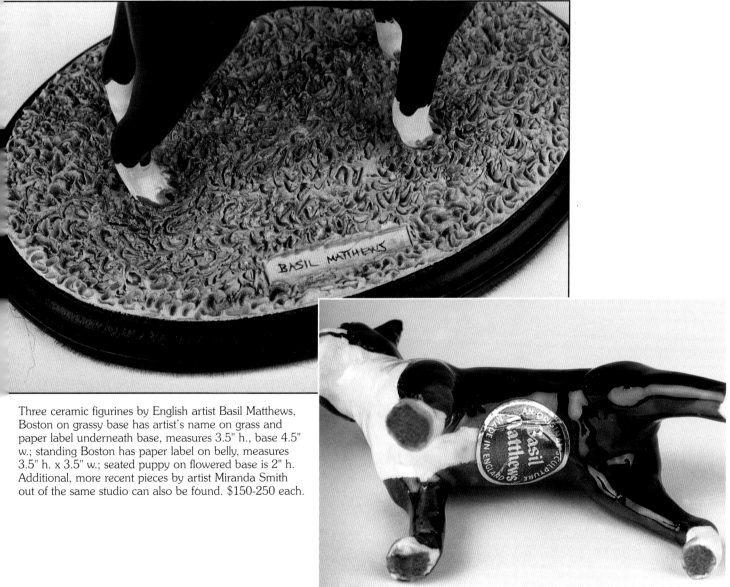

Three ceramic figurines by English artist Basil Matthews, Boston on grassy base has artist's name on grass and paper label underneath base, measures 3.5" h., base 4.5" w.; standing Boston has paper label on belly, measures 3.5" h. x 3.5" w.; seated puppy on flowered base is 2" h. Additional, more recent pieces by artist Miranda Smith out of the same studio can also be found. $150-250 each.

Ceramic figurine, sticker on bottom reads "Sylvac Made in England," 4.75" h. x 5" l. $75-125.

Two standing figurines in different colors, label on side identifies them as "Coopercraft Made in England," 7" h. x 7" w. The Coopercraft company went out of business in the early twenty-first century. $75-175 each.

Two black and white standing figurines, highly glazed, larger one marked "Made in England" on the bottom, 4" h. x 4" w. and 2.5" h. x 2.5" w. $75-150 for pair.

Set of three figurines by Hagen-Renaker, "Roland, Chip, and Chester." Roland is the standing Boston; he was only made in 1955 and is very hard to find, has a Hagen-Renaker sticker on his back, and measures 4" x 4". $300-500. The seated dogs are also from the 1950s and measure 2" and 2.5" h. $100-250 each.

Assortment of very collectible figurines from
Mortens Studio, different sizes and positions,
heights range from 5" to 3". The small standing
Boston in the center is called "Best in Show."
Puppy: $75; adults: $250+ each.

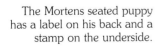

Label from underneath one of the standing Mortens figurines.

The Mortens seated puppy
has a label on his back and a
stamp on the underside.

Standing and seated ceramic figurines, both with paper labels on bottom reading "Jane Callender, California," c. 1950s, 3.5" and 2.25" h. Highly collectible. $150-300 for pair.

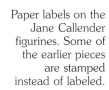

Paper labels on the Jane Callender figurines. Some of the earlier pieces are stamped instead of labeled.

Two pottery figurines by American artist Chester Nicodemus, c. 1960s-1970s. The smaller one in brown is 2.25" l. x 1.5" h., $150-250. Larger one in white measures 4" h., $250-350. Nicodemus created many different animal figurines; his works are highly collectible and sought after.

Standing figurine, made by the California based company of Robert Simmons Ceramics, c. 1940s-1960s, label on side reads "Joy 123," 4" h. $50-75.

Rookwood Pottery reclining figurines in white and brown by artist Louise Abel, Roman numeral mark underneath of XXXIV (1934), 2.25" h. x 5" l. Rare. $500+ each.

Rookwood Pottery seated figurines in white and brown, also by artist Louise Abel, white one has Roman numeral mark underneath of XXXVI (1936), brown one has XXXV (1935) underneath, 4.5" h. Very rare. $600-900 each.

Three adult Boston figurines, all Japanese,
standing dog is 5.5" h. x 6.5" w., seated dogs are
7.5" h. and 5.5" h. $25-50 each.

Assortment of Boston puppies, also Japanese, different styles and
positions, 3.5" average height. $10-30 each.

Ceramic figurine on brown and green rectangular base, by American artist Carol Moreland Marshall, artist's initials "CM" in one corner, c. 1970s, 4.5" h., base 4.5" x 3". $150-250.

Two large seated Bostons, left is bank; right is figurine only, contemporary, 8.5" h. $50-100 each.

Ceramic figurine of Boston Terrier with paw up, zig zag or "flock-ing" design on body and head, similar pose as bookends shown on page 17, American, 4.5" h. Handwritten notation on bottom identifies this piece as a Christmas gift from 1965. $25-30.

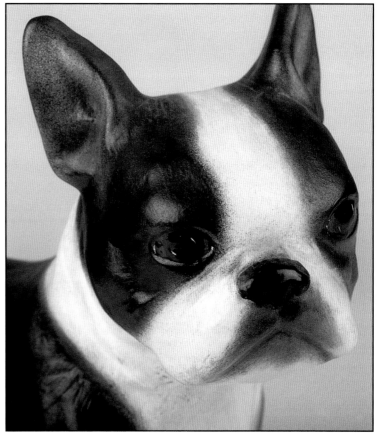

Large figurine with slightly textured matte finish, 9" h. x 11" l. $40-80.

Large ceramic seated figurine, contemporary, 10" h. $15-20.

Two large seated Bostons, similar in position and expression but one matte and one glossy, 9" and 8.5" h. $20-30 each.

Life-size ceramic figurine with one paw up, probably used as a display piece, c. 1960s, 18" h. x 14" l. Hard to find. $300-450.

Life-size ceramic figurine, lots of white on chest and front legs, signed "Townshend" on right haunch, 17" h. x 13" l. $200-250.

Three member ceramic Boston Terrier band, bandleader and two musicians,
orange jackets, Japanese, c. 1950s, 3" h. each. $100-150 for set.

Five member ceramic band, bandleader plus four musicians,
orange jackets, Japanese, c. 1950s, 2.5" h. $150-250 for set.

Diminutive Boston Terrier band, five ceramic band members wearing black jackets, German, c. 1950s, 2.5" h. each. $200-400 for set.

Even smaller five member ceramic band, also with black jackets, German, c. 1950s, 1.5" h. each. $200-400 for set.

Ceramic Boston Terrier family, mother and two pups on chain, American, mother 6.5" h., pups 2" h. $50-80.

Ceramic mother with two pups on chain, American, mother 1.75" h., pups 1" h. $35-50.

Whimsical ceramic "bug eye" mother with two pups on chain, c. 1970s, mother 5.5" h., pups 2.5" h. $30-50.

Ceramic mother with two pups on chain, mother dog is bank (opening at rear of neck), mother 8" h., pups 2.75" h. $50-75.

Ceramic mother and two pups on chain, probably Japanese, mother 6.5" h., pups 2" h. $30-60.

Ceramic mother with pup in mouth and three more pups in basket, circa 1980s, mother 4" h. $100-150.

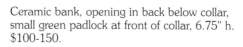

Ceramic bank, opening in back below collar, small green padlock at front of collar, 6.75" h. $100-150.

Ceramic Boston Terrier pup,
jaunty red hat and blue bow,
paper label on back reads
"An Enterprise Exclusive,
Toronto, Canada," 4" h.
$25-50.

Unusual ceramic figurine of little boy with Boston peeking around his ankles,
American, made by the Gort China Company, marked on bottom "bone
china," "Gort," and "Butch," c. 1970s, 4.5" h. $200-300. Another figurine by
Gort (not shown) features a Boston puppy seated next to a basin with towel
and soap at hand. It is hard to find and valued at $300-500.

Other Materials

Two Boston Terrier figurines made out of hydrostone (a stone-like cement), by American artist Jan Allen and signed on back, lying pup is 3" h., sitting pup is 4" h. Jan Allen worked for the Contemporary Arts of Boston from 1939 to 1957; her work is highly esteemed by collectors. $150-225 each.

A third figurine from the Jan Allen set above is lying down with head tilted up. $150-225.

Three additional Boston figurines by Jan Allen, one with bronze coating, one in black and white, and one in brindle and white, all very rare, 5.75" h. x 6" l. $500+ each.

Trio of seated Bostons sculpted out of resin, largest is 4.75" h., medium is 4.5" h., smallest is 2.75" h. $30-50 each.

Standing Boston sculpture by artist Ron Hevener of Manheim, Pennsylvania, signed "Hevener" on back of right leg and marked "No. 523" on belly, 5" h. x 5" w. $50-100.

Very heavy contemporary head study of Boston Terrier surrounded by roses, also by Ron Hevener, marked on bottom "The Bostonian," Hevener, No. 048/1000, 6.5" h. $50-100.

Resin nodder of reclining Boston Terrier, 5" h. x 3.5" l. $20-30.

Assortment of figurines, various colors and styles, some resin, heights range from 8" to 4". The figurines at bottom left and top right are by Jan Allen and are also shown on page 89, $500+ each; the center right figurine is by Northlight, $50-150; the bottom right figurine is by Donna Finnegan, a contemporary Canadian artist, $200-300; and the other two are by unknown manufacturers, $25 and up.

Three carved wooden Boston Terriers, similar in style but not a set, largest is 3.5" h., medium is 2" h., and smallest is 1.25" h. The medium size dog has tiny glass eyes. Numerous other wooden Boston figurines are known as well. $25-50 each.

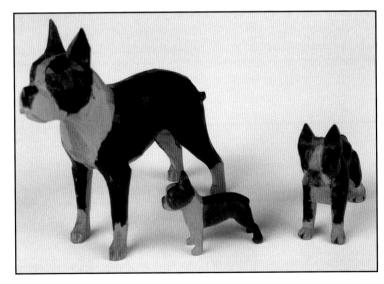

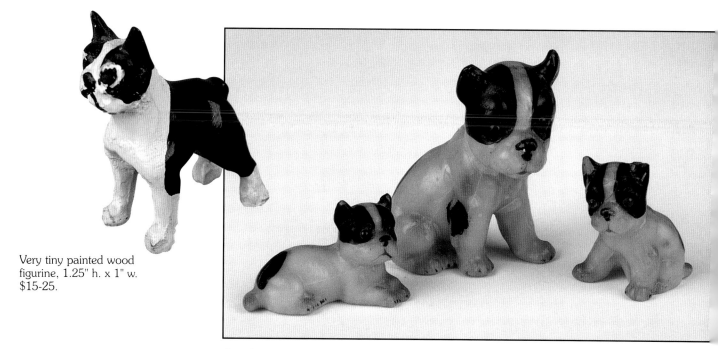

Very tiny painted wood figurine, 1.25" h. x 1" w. $15-25.

Set of mother and two pups made out of soap, American, c. 1970s-1980s, mother is 2.75" h., pups are 1.5" and 1" h. $75-100.

Chalkware figurine of standing Boston with red and gold collar, 4" h. x 3.75" w. $35-40.

Celluloid figurine, red collar around neck, American, 3" h. $25-35.

Chapter 8

Miscellany

Assortment of glass Boston Terrier heads in different colors used as paper-weights, all made by Boyd's Glass, c. 1950s-present, 3.5" h. $15-75 each.

Glass powder jar with lid in the shape of a Boston Terrier head, very heavy, American, c. 1970s, 5.5" h. Hard to find. $400-600. A very similar powder jar was made in France in the 1930s and is valued at $600-900. The French powder jar differs from the American in that it has a small silver collar with studs.

Assortment of crystal items with silver Boston Terrier heads, Austrian, c. 1930s-1950s, height range 3.5" to 7.5". From left are salt and pepper shakers, a candy container, inkwell, and personal decanter. $500-1,000+ each.

Additional crystal pieces with silver Boston Terrier heads, also Austrian, c. 1930s-1950s, height range 5.5" to 8.25". This group includes a perfume atomizer and two decanters. $500-1,000+ each.

Two figural glass candy dishes, one clear and one lavender, clear one has lid, made by American Flint Glass Co., c. 1929, 6" h. x 7.25" l. Several other colors of these candy dishes are also available. $150-250 each.

Two additional candy containers, originally filled with candy and had cardboard in bottom, American, probably made by Avon, 3.75" h. x 3" w. $60-100 each.

Hollow candy container, brown and white Boston Terrier figurine covered with a knit or jersey-like flocking, German, c. 1920s, 12" h. x 14" l. Very rare. $800-1,000.

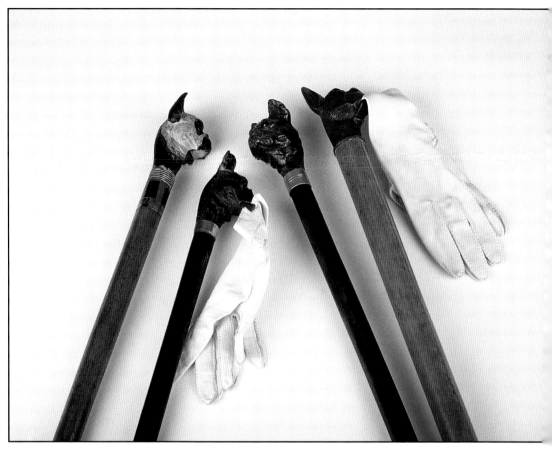

Group of four canes with Boston Terrier heads, two made to hold gloves in the mouth, all have glass eyes. $500-$1,000+ each.

Another group of canes with Boston heads. From left: painted carved wood automaton, tongue sticks out when button is pushed; large carved wooden head; horn (possibly rhinoceros?) head; ivory head. $500-$1,000+ each.

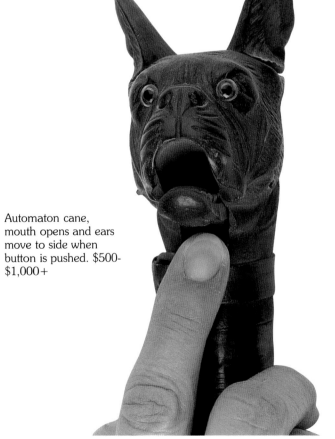

Automaton cane, mouth opens and ears move to side when button is pushed. $500-$1,000+

Wall hanging with Boston Terrier leaping over the moon, made by Plaster
Baci Studio, contemporary, 6.5" h. x 9.5" l. $25-50.

"MORTENS 3-D HEAD STUDIES"

ROYAL DESIGN wall decora
tions by famed sculptor OSCAR
MORTENS is a complete collection of different breeds and
characters. The exclusive re-
cessed hook permits simple and
sure hanging flush with wall

The Mortens Studio — Chicago 22, Ill.

Two Mortens Studio wall hangings of very expressive Boston
Terriers, label on back identifies them as part of the "Mortens 3-D
Head Studies" by sculptor Oscar Mortens, 6" h. $400-800 for pair.

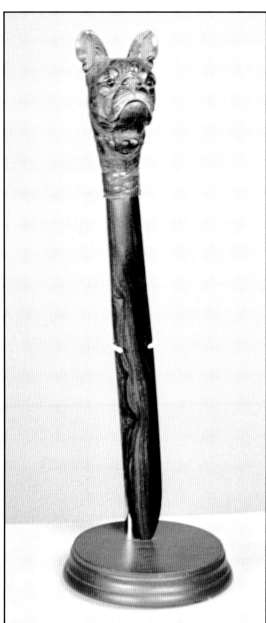

Hand carved Austrian letter opener with mechanical head, mouth opens when lever under chin is pushed, c. 1920s. Very rare. $800-$1,000.

Ceramic perfume lamp, brown and white Boston seated on white base (also found in black and white), holes in collar and on top of head for scent, plugs in from back, German, 9" h. This lamp is one of at least five known perfume lamps featuring Boston Terriers. Rare. $600-$1,000+

Art Deco porcelain book rest of Boston Terrier or French Bulldog, made by Goebel, c. 1920s-1930s, 5.5" h. x 7" l. $300-400.

Porcelain bottle stopper by Goebel, c. 1930s-1940s. Rare. $350-600.

Goebel decanter, stamped Germany in green on the bottom, 8.25" h. $125-175.

Contemporary Boston Terrier RV cookie jar and teapot, made by artist Pat Bartlett; cookie jar personalized for Tru-Mark Bostons. The artist uses 14K white gold for the bumpers and enamel for the headlights, tail lights, and hearts. No molds are used for the dogs; they are each one-of-a-kind. Cookie jar 11" h. x 6.5" w.; teapot 9" h. x 10.5" wide. $200-250 each.

Napco sticker on bottom of the planter.

Ceramic planter made by Napco, Boston Terrier figurine on front, whole planter is 4" h. x 5" w. $30-50.

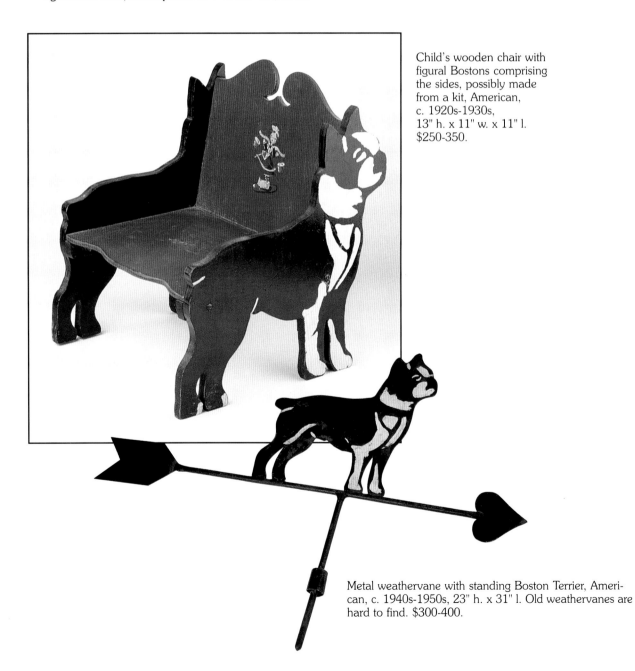

Child's wooden chair with figural Bostons comprising the sides, possibly made from a kit, American, c. 1920s-1930s, 13" h. x 11" w. x 11" l. $250-350.

Metal weathervane with standing Boston Terrier, American, c. 1940s-1950s, 23" h. x 31" l. Old weathervanes are hard to find. $300-400.

Pair of cast iron fireplace andirons with glass eyes, made by Howe, c. 1920s, 17" h. $600-800 for pair.

This painted version of the Howe fireplace andirons is especially rare. They also have glass eyes and are c. 1920s, 17" h. $1,500-1,800 for pair.

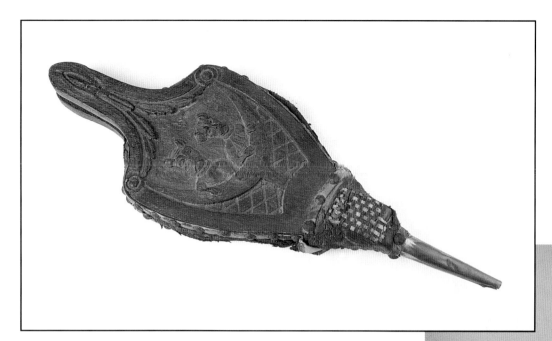

Fireplace bellows, Boston Terrier and Jack Russell Terrier on front, American, c. 1920s, 22" l. $100-150.

Ceramic, Boston Terrier music box/ decanter, a Hoffman Original that plays "How Much is That Doggie in the Window," 6" h. x 4" l. $75-150.

Label from the side of the music box/ decanter shows that it is filled with Kentucky Straight Bourbon Whiskey.

Ceramic music box by artist Helen Johnson, old fashioned Victrola adorned with Boston Terrier musicians and dancers, contemporary, 7.5" h. $50-75.

Carousel music box by Dannyquest Designs, ceramic Boston figurine on wooden base, plays "You Light Up My Life," c. 1990s, 11" h. x 10" l. $100-125.

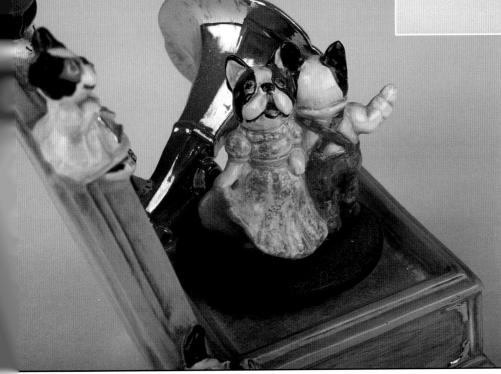

Detail of dancers on the Victrola music box.

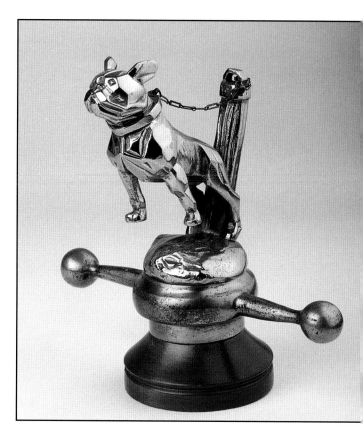

Bronze radiator mascot of Boston Terrier or French Bulldog, originally screwed onto exposed radiator cap, French, c. 1920s, figurine only is 7" h. Although many radiator mascots as well as other pieces were made by the Mack Trucking Co. and depict English Bulldogs (the Mack company mascot), this radiator mascot was not made by Mack. $1,000+

Vienna bronze candelabra with figurine of Boston Terrier or French Bulldog, c. 1910s-1920s, 7" h. $1,000+

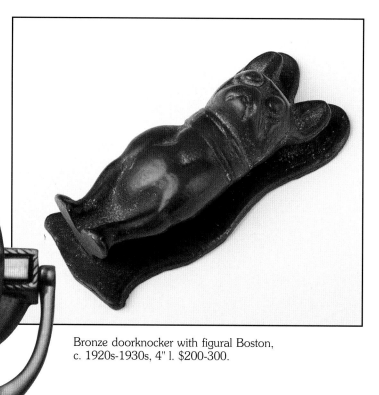

Bronze doorknocker with figural Boston, c. 1920s-1930s, 4" l. $200-300.

Brass doorknocker from England, c. 1980s-1990s. Hard to find. $250-350.

Light bulb with Boston Terrier filament that lights up in red when bulb is plugged in, c. 1950s-1960s. $250-350.

Hand carved wooden nutcracker with Boston Terrier head, probably American, 7.75" h. Hard to find. $300-400.

Set of sterling silver nut trays, ashtray, and match box holder, all featuring the same standing Boston design in the center, c. 1920s. Large tray is 3.75" dia., smaller ones are 2.75" dia. $300-400 for set.

Sterling silver jigger with Boston Terrier or possibly pug design, goes over liquor bottle, inscribed on bottom: "TWS from HB, Feb 6 1932 – 21 yrs.," 2.5" h. $300-500.

Vienna bronze box with two Boston heads in relief on top, c. 1920s-1930s, 2.5" x 4" x 3". Rare. $600+

Printer's blocks in assorted sizes and designs, all featuring Boston Terriers. $75-150 for assortment.

Assortment of cast iron party favors in different positions, c. 1930s, approximately 1.5" h. x 1.5" w. each. Some have a slit in the mouth to accommodate a placecard. $400-500 for the group.

Ceramic vignette on wooden base with Boston Terrier and feline pal, "Hittin the Trail" by American artist Lowell Davis, Border Fine Arts, 4" h. Hard to find. $300-500.

Ceramic western vignette by artist Helen Johnson, Boston Terrier family
around campfire with covered wagon and bones, contemporary, 2" h. $25-50.

Bronze figurine of Boston Terrier with pipe inside small
glass bell jar, 2" h. $200-300.

Pair of miniature ceramic one-of-a-kind figurines, man and woman with Bostons, made by artist Cece Canaga, c. 1980s, man is 3" h., woman is 2.5" h. $400-500 for pair.

Victorian ladies with Bostons, also made by Cece Canaga, c. 1980s, one-of-a-kind pieces, 5" h. $600-800 for pair.

Handmade Christmas vignette modeled after a Norman Rockwell painting, made by Mary Margaret Logan, contemporary, 8" h.

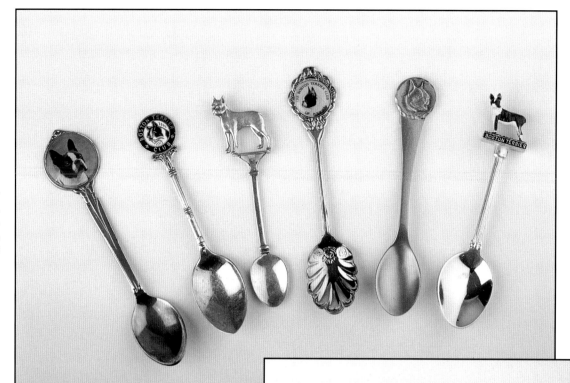

Assortment of silver baby spoons and souvenir spoons, each incorporating a Boston Terrier design in the handle. $35-150 each, higher for sterling.

Metal toothpick holder, souvenir of Niagara Falls with Boston Terrier or French Bulldog on side, 1.5" h. $40-60.

Assortment of silver medallions from various dog clubs, some inscribed on the back, c. 1910s-1930s. $500-800 for the group.

Old dog club medal awarded by the Boston Terrier Club of America for Best of Breed, 2.5" dia. $150-250.

Figural pin of Boston Terrier head, label on back reads "Top Dog Art, © 1993, Handpainted," 2" l. $30-40.

Necklace featuring Boston Terrier head, made by artist Trophy Frederick. $75-100.

Cloissoné enameled pins, standing Boston Terriers, 1.25" h. $75-100 for both.

Sterling silver pin, running Boston Terrier, 3" l. $30-40.

Chapter 9

Personal Items and Novelties

Schuco compact, very unusual and hard to find, German, c. 1920s. $800-$1,000.

Ladies wooden powder puff holder, egg shaped and with intricate carved detail on outside, carved bone or ivory dog with powder puff inside, American, c. 1950s, 2" h. $200-300.

Two ladies compacts with Boston Terrier designs, both c. 1930s. Round compact is all enamel, European, 2" dia., very rare. $500. Rectangular compact is possibly by Evans, 3.25" x 2.25". $200-300.

Two additional compacts, both c. 1930s. Round is 2" dia.; square is by Evans, 2" sq. $150-225 each.

Set of assorted items all with different functions but featuring the same Boston Terrier figurine, possibly made of gutta percha (a tough, rubberlike material made from the dried sap of Malaysian sapodilla trees), c. 1940s-1950s, average height 3.5". Set includes a brush, napkin ring, perpetual calendar, bottle stopper, corkscrew, tape measure, shoe horn, bridge pencil holder, bottle opener, pencil holder. $1,000-1,200 for set.

Group of three brushes with Boston Terrier heads, part of a collection of twelve, heads made of celluloid (left) and ceramic (center and right), c. 1940s-1950s, 7" l. $250-350 for set.

Wooden shoe brush, American, 3.75" long. $100-150.

Parian bonbonniere in the shape of a Boston Terrier head, English, c. 1920s, 1.75" h. Bonbonnieres were small containers for candy, often carried by ladies in their purses. Rare. $600-900.

Left and right:
Assortment of reverse intaglio or Essex crystal pieces, c. 1930s-1950s. Includes cigarette holder, matchbook holder, letter clip, stickpin, square boxes, tape measure, pill box, letter clip, trump indicator, corkscrew, stamp box, cigarette case, perfume container, rectangular box, paper holder. Rare. $1,200-1,500+ for group.

Detail of reverse intaglio crystal from gold colored square box.

Porcelain tape measure, Boston Terrier head, c. 1930s-1940s. Porcelain head tape measures such as this were very popular during the 1930s and 40s and are hard to find. $300-400.

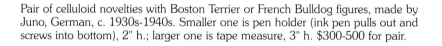

Pair of celluloid novelties with Boston Terrier or French Bulldog figures, made by Juno, German, c. 1930s-1940s. Smaller one is pen holder (ink pen pulls out and screws into bottom), 2" h.; larger one is tape measure, 3" h. $300-500 for pair.

Pair of celluloid pencil sharpeners, both the same except for different color on band at bottom, probably German, c. 1930s-1940s, 1.25" h. $500-600 for pair.

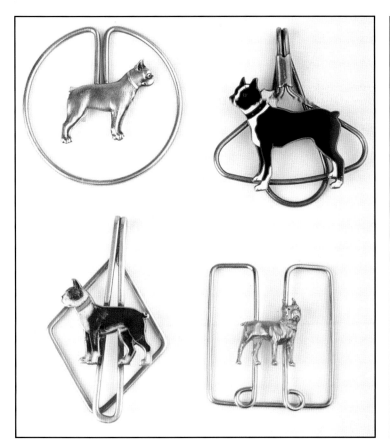

Assortment of sterling silver and enamel money clips with figural Boston Terrier designs. $75-125 each.

Trio of sterling silver novelties made by Unger Bros., c. 1920s-1930s, each decorated with tiny Boston Terrier heads. Clockwise, from top: bottle stopper, 2" h., $100-200; mini letter opener, 4.5" l., $300-500; whistle, 1.75" l., $300-500.

Pair of sterling silver place card holders in the shape of Boston Terriers, American, 1" h. each. $125-250 for pair.

Small sterling triangular perfume holder with Boston head on front, c. 1930s, 2" long. $400-600.

Perfume holder with Lalique frosted crystal Boston Terrier or French Bulldog on top, perfume "Toujours Fidele" by D'Orsay, c. 1920s, 3.25" h. $400-600.

Chapter 10

Plates and Plaques

1910 calendar plate with Boston Terrier,
7.5" dia. $100-150.

Marble plaque with original painting of
"Bunny" by contemporary American artist
Sherry Morgan, signed on the front, 8" h. x
8.5" w. $100-150.

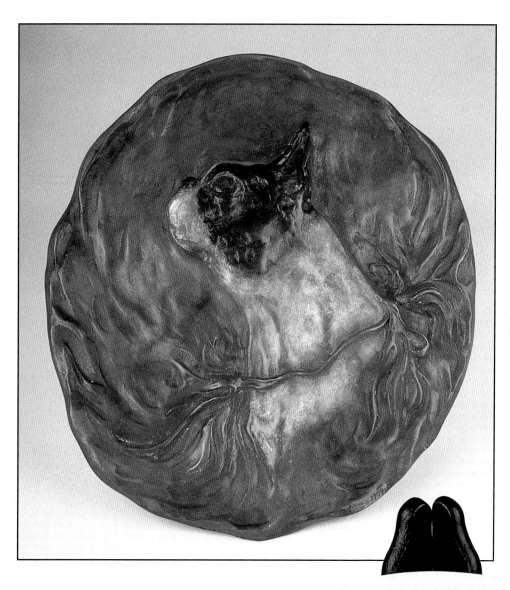

Elegant bronze plaque made by Kathleen O'Bryan Hedges, c. 1990s, 10" dia. Two figural sculptures by this American artist are shown on page 52. $200-300.

Rare Dresden plate with three Boston Terrier puppies, c. 1940s-1950s, 4.5" h. x 6" l. $1,000+

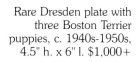

Pair of hand painted ceramic plates with Boston pups, made by California artist Diana Orange, c. 1980s, 9.5" dia. each. $250-300 for pair.

Bronze plaque with Boston Terrier head and dog's name "Goggles" on the bottom, signed below head is "RM Mason, 1929," 6.75" dia. $1,000-1,200.

Detail of plaque, showing artist's signature and "Goggles" at the bottom.

Handpainted ceramic plate by Nippon, head and upper body of Boston,
"blown out" design in slight relief, c. 1920s, 10.5" dia. $800-$1,000.

Ceramic plate, Boston puppy with flowers, American, contemporary, 7.5" dia. $25-35.

Small ceramic plate with reclining Boston puppy, American,
contemporary, 6" dia. $25-35.

These five limited edition china plates, part of a collection of nineteen, are from Laurelwood, which has been producing dog breed collector plates since 1977. Only 150 plates of each design are made, each in a sepia tone design on an ivory colored plate with a handpainted gold rim. These five are all signed on the front by artist Paula Zan and measure 8.5" in diameter. Back of the plate above reads: "Boston terrier, Limited Edition of 150, Laurelwood, Lehigh Valley, Pa 18001, © Laurelwood 1985." $35-150 each.

1993 Laurelwood plate.

1987 Laurelwood plate.

1989 Laurelwood plate.

1990 Laurelwood plate.

Boston Terrier Rescue Net

Boston Terrier Rescue Net was formed in 199_ by a small group of dedicated Boston Terrier fanciers, with the mission to raise money to support the rescue of this wonderful breed. Through their various fundraising efforts, this not-for-profit group is able to aid in the medical and other expenses of rescued Bostons, and in the location of adoptive and foster homes for Bostons that have been abandoned, abused, left homeless or simply no longer wanted by their owners.

Boston Terrier Rescue Net is pleased to feature the winner of our "Caring Hands, Loving Hearts" photo contest on this collector plate. Adorable "Chris" was entered by Jim and Dot Leverett of Georgetown, Tennessee.

Illustration by V. Susan Miller, 2000

Second in a Series

Limited Edition 48 of 250

www.bostonrescue.net

Limited edition collector plate signed by artist Vera Susan Miller, © 2000, #48 of 250, second in a series produced by the Boston Terrier Rescue Net, illustration based on winner of the "Caring Hands, Loving Hearts" photo contest, 8" dia. $25-30.

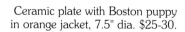

Ceramic plate with Boston puppy in orange jacket, 7.5" dia. $25-30.

Chapter 11

Products and Promotions

Boston Shoe Cream
container, c. 1950s, 2.5"
dia. $35-50.

A trio of cans for diverse products, all using a Boston
Terrier image on the front: Hughes Shampoo, 4" h.;
Hammond's Grape Dust for Roses, 5.5" h.; Barking Dog
Smoking Mixture, 4.25" h. $200-350 for group.

Carton and individual
package of Barking Dog
Cigarettes, made by Philip
Morris, c. 1930s-1940s. Very
hard to find. $300-400 for
both.

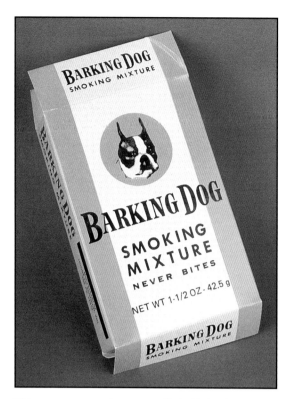

This contemporary box originally held 1-1/2 oz. of Barking Dog Smoking Mixture. $25-50.

This fire extinguisher is labeled "Bull Dog Brand" but features a Boston Terrier logo at the top, American, 22" l. $250-450.

Barking Dog Cigarettes display cabinet, used to hold
Barking Dog products in general stores or similar
locations, 21.5" x 14" x 6.5". Rare. $600-800.

Bank advertising Rival Dog Food, Boston Terrier in top right corner, 2.5" h. $15-25.

Small paper bag and large burlap sack, both for Old Mike brand of potatoes and featuring Boston Terrier as logo. $150-200 for both.

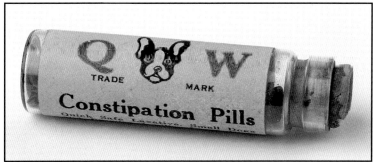

Q-W Constipation Pills, Boston Terrier design on label, could be used for dogs as well as humans, c. 1920s-1930s. $25-50.

Small paper advertising bag for Sutrite Hosiety, c. 1930s-1940s. $30-50.

Stuffed dog symbolizing Boston University's mascot, c. 1950s-1960s, 3.5" x 4". $25-35.

Large reclining Boston Terrier made of papier mâché, advertising piece for Bryant Heating with "The Bryant Pup" in gold letters on red collar, 10.5" h. x 20" l. $400-800.

Another large advertising piece for Bryant Heating, also made of papier mâché but in a different pose, 10.5" h. x 20" l. $400-800.

Ceramic advertising figurine for Bryant Heating, much smaller than the papier mâché dogs, 2.5" h. x 4" w. $35-50.

Mechanical pencil with Boston Terrier design, advertising piece
for Milk-Bone Dog Biscuit, c. 1940s. $150-200.

Assortment of advertising match covers, c. 1910 to late twentieth century, two from Boston University and all
featuring a Boston Terrier. Many additional match covers with Bostons are available to collectors. $1-$25 each.

Assortment of pinback buttons, all with Boston Terrier designs. $75-150 each.

Tiny pinback button, reads "Dr. A.C. Daniels, Dog Remedies," .75" dia. Originally founded in 1878 in Boston, the Dr. A.C. Daniels Company sold products for horses, dogs, cats, sheep, cows, and poultry. Vintage promotional items and product containers from this company (still in business today) are very popular with collectors. $250-300.

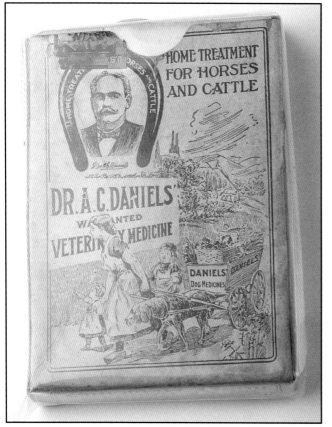

Deck of cards advertising Dr. A.C. Daniels veterinary products c. 1930s, illustration on cards inside is of Boston Terrier. Rare. $600-900.

Three cardboard candy boxes decorated with Boston Terrier themes, c. 1920s-1930s. Center and right boxes are from Lowneys's; back of right box reads: "No. 524 Two Pounds Assorted Chocolates, Made in U. S. America by The Walter M. Lowney Company, Boston, Mass." Small box, $50-75; large boxes, $75-125 each.

Tin over cardboard sign for El Bart Dry Gin, c. 1910s-1920s, 13" x 9". Rare. $1,200-1,500+

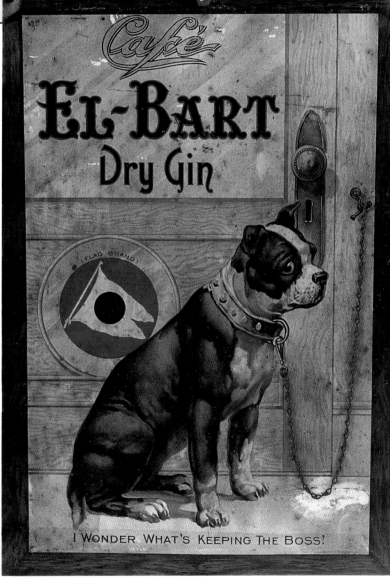

Die cut tin sign for Cortez Cigars, c. 1916, 16" x 12". Rare. $1,000-1,500.

Fruit crate label for Security Brand oranges with image of Boston, 9.5" x 10.5". $500+

Wadsworth Whiskey advertising sign, c. 1910s-1920s, 21" x 15.5". Rare. $1,200-1,500+

These four magazine advertisements, all featuring Boston Terriers, represent just a sampling of the many ads available to collectors. $5-25 each.

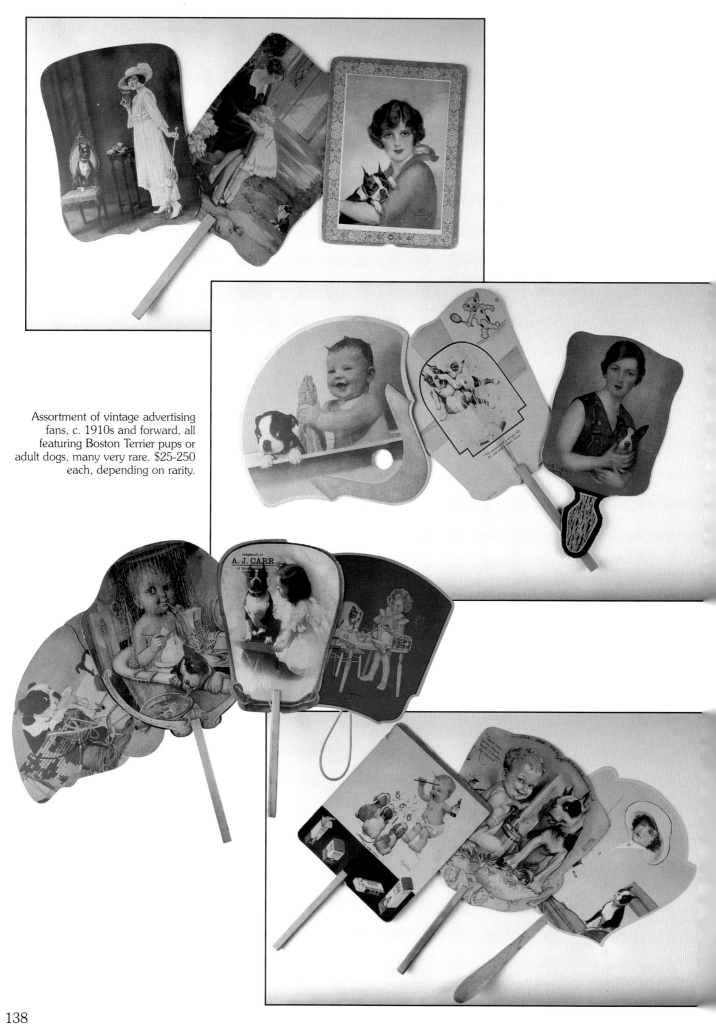

Assortment of vintage advertising fans, c. 1910s and forward, all featuring Boston Terrier pups or adult dogs, many very rare. $25-250 each, depending on rarity.

Chapter 12

Smoking Accessories

Ashtray with Vienna bronze figurine of Boston Terrier on green marble base, 3.5" h. x 3.5" w. $400-600.

Brass smoke set by Bradley & Hubbard, includes humidor, match holder, and ashtray all decorated with Boston Terrier heads in relief, c. 1920s-1930s, 15" h. x 14" dia. $1,200-1,500.

Bronze finish ashtray with face-to-face Boston and bulldog figurines, 3.5" h. x 7" w. $200-300.

Metal ashtray with sleeping Boston Terrier figurine and attached match holder, 3" h. x 4.25" w. $100-200.

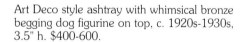

Art Deco style ashtray with whimsical bronze begging dog figurine on top, c. 1920s-1930s, 3.5" h. $400-600.

Onyx ashtray with two Vienna bronze Bostons, c. 1920-1930s, 2" h. x 4" dia. $600-800.

Ashtray with Vienna bronze figurine of Boston Terrier on onyx base, figurine 2.25" h. x 2.25" w., base 4" x 5". $600-800.

Brown and white cast iron cigarette lighter, 5" h. x 5.5" l. A similar non-lighter figurine is shown on page 47. $250-450.

Detail, showing cigarette lighter end.

Ceramic cigarette lighter, Boston sitting next to lamp marked "Gatineau, Que.," 5.5" h. $100-200.

Bradley & Hubbard match holder with Boston Terrier face in relief on front and back, c. 1930s, 2.75" h. x 1.75" w. $300-450.

Metal cigarette lighter, Boston Terrier or French Bulldog, head lifts up to access lighter, 2.5" h. x 3" l. $100-200.

Pair of sterling silver matchsafes, both monogrammed on back, c. 1920s, 2.5" h. $400-500 each.

Two matchboxes with Boston Terrier motifs, top has reverse intaglio crystal design in center, each approximately .5" h. and 1.75" l. $200-300 each.

Left: Bronze cigar cutter, Boston Terrier or French Bulldog on end, c. 1920s, 6.5" long. Right: desktop cigar cutter, painted bronze with Boston Terrier or French Bulldog on top, c. 1920s-1930s, 4.5" h., 6.5" x 4.25" base. $500+ each.

Ceramic Boston Terrier head ashtray, Japanese, 2.5" h. $25-50.

Metal pipe holder, Boston Terrier or French Bulldog, 3" h. $75-125.

Matching cigarette case and lighter decorated with Boston heads, probably silverplate with enameling, c. 1940s. Case, 3" x 3.5"; lighter, 2" h. $200-400 each.

Cigarette case with very expressive Boston on front, sterling silver and European enamel, German, possibly from the 1920s, 3" x 3". $800-1,000.

Chapter 13

Tableware

Ceramic salt and pepper shakers, 2.5" h., common. $25-30. Salt and pepper shakers are very popular collectibles and quite a few feature Boston Terriers. Those shown here are representative of the many sets available.

Ceramic salt and pepper shakers, 3" h., common. $25-30.

Ceramic salt and pepper shakers, marked Japan, two Bostons getting into trouble! 2" h. $25-30.

Ceramic "bug eye" salt and pepper shakers, 3.5" h. $25-30.

Three sets of ceramic salt and pepper shakers. From left: accordion players, 3" h.; wearing blue coats, 2.5" h; with green top hats, 3" h. $25-30 each.

Ceramic salt and pepper shakers, wounded Boston pups, c. 1960s, 3.5" h. $25-30.

Ceramic salt and pepper shakers, 2.75" h., common. $25-30.

Ceramic salt and pepper shakers, one seated and one lying down, 3.5" h. (seated). $25-30.

Rosemeade salt and pepper shakers, Boston Terrier heads, 2.5" h. Hard to find. $200-300.

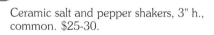

Ceramic salt and pepper shakers, 3" h., common. $25-30.

Very small ceramic salt and pepper shakers, 1.75" h. $25-30.

Carved wooden stein with Boston head in relief on front, handle made of horn, c. 1910-1920s, 8.5" h. $600-800.

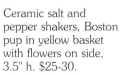

Ceramic salt and pepper shakers, Boston pup in yellow basket with flowers on side, 3.5" h. $25-30.

Ceramic Boston Terrier creamer, Japanese, c. 1970s, 3.5" h. $35-50.

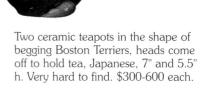

Two ceramic teapots in the shape of begging Boston Terriers, heads come off to hold tea, Japanese, 7" and 5.5" h. Very hard to find. $300-600 each.

Ceramic condiment set, mother Boston and two pups in basket, Japanese, 3.5" h. $75-125.

Ceramic gravy boat with two Bostons paddling, one-of-a-kind contemporary piece by artist Yula Anderson, 6" h. x 9" l. $75-150.

Set of six ceramic cups from Japan, possibly made for sake, each 2" h. $35-50 for set.

147

Two sugar dispensers, glass with silver heads and amber colored glass eyes, 5.5" h. $300-450 each.

Sterling silver baby's fork and spoon, each decorated with Boston Terrier on handle. $75-125 for pair.

Two Bavarian porcelain steins, both c. 1920s. Left hand stein has black and white Boston heads on either side and gold trim, 7" h.; right hand stein has Dragonwyck handle, 5.5 " h. Both very rare. $600+ each.

Chapter 14

Textiles and Needlework

Two handkerchiefs with Boston Terrier motifs; many other handkerchiefs with Bostons are also available. The white handkerchief on the left has a label for Bloch Freres, "Made in Switzerland." $20-25 each.

One-of-a-kind needlepoint rug made by Joan Hiller, c. early 1990s, 33.5" x 23".

Vintage needlepoint doorstop with Boston Terrier pup, 7" h. x 4.5" w. $75-125.

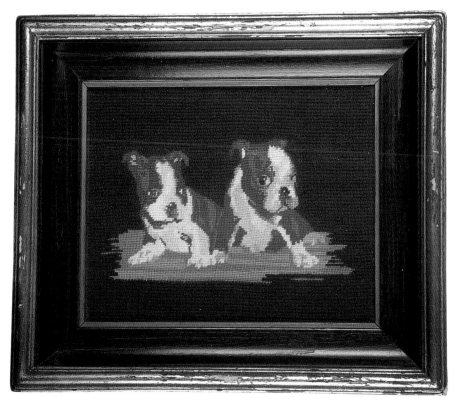

Framed needlepoint picture of two reclining
Bostons, 9.5" x 7.5". $75-150.

Framed Boston textile, wool-like texture, 17" x 19.5".$75-150.

Chapter 15

Toys and Dolls

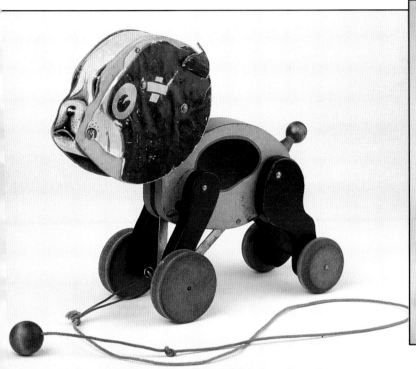

Wooden pull toy, "Growly Grouch" by All Fair Toys, American, c. 1920s, 8" h. x 9.5" l. Very hard to find. $400-600.

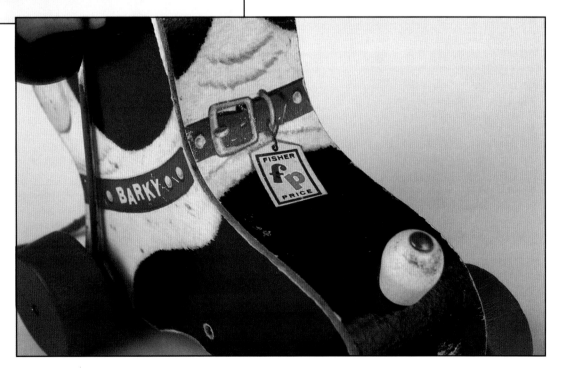

"Barky" wooden pull toy by Fisher Price, painted tag attached to collar, 4" h. x 5" l. $100-150.

Wooden pull toy with Boston playing piano, American, c. 1940s-1950s, 12" x 12". Very hard to find. $250-500.

Old unusual stuffed toy, seated Boston wearing orange cap and bow, German, c. 1910, 8.5" h. $300-400.

Toy dog, seated Boston with bisque head and plush body, unusual combination of materials, German, c. 1920-1930s, 11" h. x 8" l. $800-1,000+

Three German felt or flocked toys, Boston Terriers or French Bulldogs, two on wheels and one windup (with orange cloth in mouth), largest is 9" x 9", medium is 8" x 8", smallest is 8" x 4.5". $400-600 each.

Standing windup toy, dressed Boston Terrier, bisque head with plush body, shakes arms and legs when wound up, probably German, c. 1930s, 9" h. $500+

Steiff pull toy, Boston Terrier or French Bulldog on red wheels, possibly mohair, German, c. 1920s, 6" h. x 7.5" l. Rare. $800+

Steiff stuffed animal, Boston Terrier or French Bulldog, has original Steiff ear button ear and leather collar, German, c. 1920s, very rare especially due to bellows that are pushed for music. $1,200-1,500.

Steiff stuffed animal, Boston Terrier or French Bulldog, very unique, also has original Steiff ear button, leather collar with bell and Steiff designations, German, c. 1920s. 14" h. Very rare. $2,000+

Pair of colorful celluloid Boston Terrier windup drummers, one on wooden base, probably American, c. 1950s, 8" and 9.25" h. $300-400 each.

Schuco mechanical toy, Boston Terrier or French Bulldog, German, c. 1920s. When tail is moved one way, the head shakes "yes"; when tail is moved the other way, the head shakes "no." Very rare. $1,000-1,200.

Very large and unusual papier mâché pull toy on wheels, 16" h. x 17" l. $500-1,000+

Schuco mechanical toy made of felt, German, c. 1930s. Pushing on the dog's neck opens the mouth. Very rare. $800-1,200.

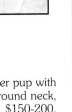

Ceramic jointed Boston Terrier pup with orange bow and small bell around neck, 2.5" h. Hard to find. $150-200.

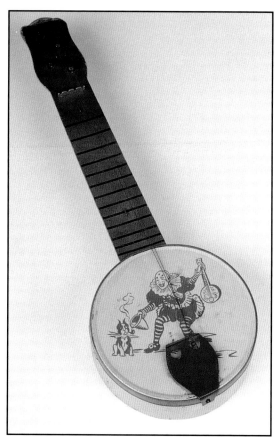

Old tin toy banjo, illustrated with clown and pipe-smoking Boston Terrier, 15" l. x 5" w. $200-400.

Tin snapper with Boston Terrier and cat design, German, c. 1930s. Front part of Boston comes up and cat comes out of chimney when toy is snapped. $300-500.

Left: Vintage tin truck with seated Boston Terrier among other dogs on side, 3" h. x 7" l. Right: Vintage tin top with Boston Terrier one of four dogs running on top, reads "Made in USA." $100-150 each.

Hand carved, one-of-a-kind wooden marionette, made in the 1970s. Strings move head, legs, and body. $500+

Set of wooden Boston Terrier nesting dolls, signed on bottom by artist Joni Clear, c. 2000, height range 2.25" to 1.75". $75-125 for set.

Miniature Boston Terrier made of hard plastic with original box, one of the "Blue Ribbon Dogs" by Marx, c. 1960s, dog is 1" h. $30-50.

Child's sand pail with Boston Terrier in decoration, American, c. 1920s, very rare. $500+

These droll-looking Bostons made of polymer clay are riding on die cast Harley Davidson scale model motorcycles with moveable parts. On one motorcycle is a dad with tiny pup in his backpack, on the other is a pregnant Boston with her companion, 4.5" h. $50-100 each.

Red metal train carrying smiling Bostons, made by artist Donna Berner, 4" h. x 11.5" l. $200-300.

Three additional Bostons by Donna Berner are riding on a hobby horse, holding balloons, and sitting in a red wagon, 5" h. (horse rider). $50-75 each.

Pair of animal head dolls, bisque Boston Terrier heads, c. 1980s, 6" and 5.75" h. Very rare. $300-500 each.

Ceramic grouping of happy Boston Terrier musicians on wooden base, also by Donna Berner, c. 1980s-1990s, each figurine 4.5" h. $150-250.

Bibliography

American Kennel Club. *The Complete Dog Book, 19th Edition, Revised* New York: Howell Book House, 1998.

Braunstein, Ethel. *The Complete Boston Terrier*. New York: Howell Book House, 1973.

Candland, Bob & Eleanor. *Boston Terriers*. Neptune City, NJ: T.F.H. Publications, Inc., 1998.

Hausman, Gerald and Loretta. *The Mythology of Dogs: Canine Legend and Lore Throughout the Ages*. New York: St. Martin's Press, 1997.

Huddleston, Arthur. R. *The Boston Terrier*. Fairfax, VA: Denlinger's Publishers, Ltd., 1985.

Leiner, Barri, and Marie Moss. *Flea Market Fidos*. New York: Stewart, Tabori & Chang, 2002.

Perry, Vincent G. *The Boston Terrier*. Chicago, IL: Judy Publishing Co., 1928.

Robak, Patricia. *Dog Antiques and Collectibles*. Atglen, PA: Schiffer Publishing Ltd., 1999.

Sheehan, Laurence. *Living with Dogs: Collecting and Traditions, At Home and Afield*. New York: Clarkson N. Potter, Inc., 1999.

"The end"

Index